SCIENTIFIC AMERICAN EDUCATIONAL PUBLISHING

DESSERT CHEMISTRY

10 FUN PROJECTS USING SWEETS

BRING SCIENCE HOME

Published in 2024 by Scientific American Educational Publishing in association with
The Rosen Publishing Group 29 East 21st Street, New York, NY 10010

Contains material from Scientific American , a division of Springer Nature America, Inc., reprinted by permission as well as original material from The Rosen Publishing Group .

First Edition

Editor: Michael Rajczak
Designer: Rachel Rising

Activity on page 5 by Science Buddies/ Megan Arnett on November 30, 2017; page 11 by Science Buddies, Megan Arnett on August 24, 2017; page 19 by Science Buddies, Sabine De Brabandere on May 11, 2017; page 23 by Science Buddies, Sabine De Brabandere on November 22, 2018; page 29 by Science Buddies, Sabine De Brabandere on July 6, 2017; page 35 by Science Buddies, Megan Arnett on July 28, 2016; page 40 by Science Buddies on March 26, 2015; page 45 by Science Buddies on April 25, 2014; page 50 By Science Buddies on December 20, 2012; page 57 By Science Buddies, Ben Finio on November 2, 2017.

All illustrations by Continuum Content Solutions

Photo Credits: pp. 3, 4, 5, 8, 16, 19, 21, 22, 26, 29, 33, 38, 40, 43, 44, 48, 55, 57, 59, 60 cve iv/ Shutterstock.com; pp. 4, 5, 11, 19, 23, 29, 35, 40, 45, 50 ,57 Anna Frajtova/Shutterstock.com.

Library of Congress Cataloging-in-Publication Data

Names: Scientific American, inc., editor. | Scientific American Educational Publishing.
Title: Dessert chemistry : 10 fun projects using sweets / Scientific American editors.
Description: Buffalo, NY : Scientific American Educational Publishing, an imprint of Rosen Publishing, [2024] | Series: Bring science home | Includes bibliographical references and index. | Audience: Grades 7-9
Identifiers: LCCN 2022058724 (print) | LCCN 2022058725 (ebook) | ISBN 9781725348981 (library binding) | ISBN 9781725348974 (paperback) | ISBN 9781725348998 (ebook)
Subjects: LCSH: Chemistry--Experiments--Juvenile literature. | Desserts--Juvenile literature.
Classification: LCC QD38 .D47 2024 (print) | LCC QD38 (ebook) | DDC 540.78--dc23/eng20230323
LC record available at https://lccn.loc.gov/2022058724
LC ebook record available at https://lccn.loc.gov/2022058725

Manufactured in the United States of America

CPSIA Compliance Information: Batch #SACS24. For further information contact Rosen Publishing, New York, New York at 1-800-237-9932.

CONTENTS

⚛ THESE ACTIVITIES INCLUDE SCIENCE FAIR PROJECT IDEAS.

INTRODUCTION

The best part of learning is when it tastes delicious. This book of 10 experiments lets you investigate and discover some very interesting concepts about desserts and other sweet treats. Imagine your parents or teachers asking why you are making a mess. You can smile and say, "This mess just happens to be a science experiment!"

Projects marked with ⚛ include a section called Science Fair Project Ideas. These ideas can help you develop your own original science fair project. Science fair judges tend to reward creative thought and imagination, and it helps if you are really interested in your project. You will also need to follow the scientific method. See page 61 for more information about that.

Invisible Marshmallow Ink

Send Sweet Secret Messages Using Marshmallows!

MAKE IT A MARSHMALLOW MAILLARD MESSAGE! WHAT WILL YOU WRITE?

When we put toast in the toaster or add marshmallows to the top of our sweet potatoes for a baked holiday dish. we expect them to turn brown and to develop a sweet, caramelized flavor. Although we expect it to happen, do you know why certain foods take on these new colors and flavors as they are toasted? In this activity you will explore the reaction that creates these tasty, toasty treats and experiment with speeding up and slowing down the process.

PROJECT TIME

30 to 40 minutes

KEY CONCEPTS

Chemistry
Acids
Bases
Maillard reactions

BACKGROUND

The process of toasting foods that we enjoy takes place because of a phenomenon known as the Maillard reaction (named after its discoverer, Louis-Camille Maillard). This reaction is actually a sequence of multiple reactions that take place between proteins and sugars when we heat our food. When the reactive carbonyl group of the sugars in the food react with the amino group of the protein, a series of new molecules are created. These new molecules contain hundreds of different flavor and aroma compounds that give rise to the caramel flavors and scents we associate with toasting.

Foods can turn brown for a number of reasons that happen outside of the Maillard reaction. When chicken turns black on the grill or when sliced fruit begins to brown when left out, other chemical reactions are at work. The Maillard reaction is important for baking because it turns our cookies and breads that delicious golden color that we've come to expect.

In this activity you'll be exploring the Maillard reaction—and how you can speed it up and slow it down with some simple kitchen chemistry!

MATERIALS

- Marshmallows (at least 10)
- Q-tips
- Toaster oven or regular oven
- Baking soda
- Milk
- Lemon juice
- Vinegar
- Water
- Five small paper cups
- Paper and pen or pencil
- Measuring spoon (1 tablespoon, about 15 ml)
- Permanent marker
- Baking tray (or tray for toaster oven)
- Tinfoil
- An adult helper

PREPARATION

- Use your marker to number the paper cups with 1-5.
- Create a table that lists "Number on Cup" in the left column, with each cell below labeled 1, 2, 3, 4 and 5, respectively, and "Liquid in Cup" in the right column, with each of the following liquids listed in this order: water, baking soda, vinegar, milk, and lemon juice. So 1 should line up with water, 2 with baking soda, and so on.
- Add at least 3 tablespoons (about 44 ml) of each of your liquid ingredients to the designated cup, using the table as a reference. For the baking soda, add 1 tablespoon (about 15 ml) of baking soda to 5 tablespoons (74 ml) of water and stir.
- Place a clean Q-tip in each cup.
- Cover your baking tray in tinfoil.
- Arrange your 10 marshmallows on your baking tray in two rows, five marshmallows across.

PROCEDURE

- On the first column of two marshmallows, use your Q-tip dipped in water to paint the number 1 on each marshmallow.
- On the second column of two marshmallows, use your Q-tip dipped in baking soda to paint the number 2 on each marshmallow.
- On the third column of two marshmallows, use your Q-tip dipped in vinegar to paint the number 3 on each marshmallow.
- On the fourth column of two marshmallows, use your Q-tip dipped in milk to paint the number 4 on each marshmallow.

- On the fifth column of two marshmallows, use your Q-tip dipped in lemon juice to paint the number 5 on each marshmallow.
- Ask your adult helper to carefully place your tray in the oven on the top rack. Have your adult helper set the temperature to broil. *Do you have any guesses as to what you might see on the marshmallows painted with different liquids?*
- Monitor your marshmallows constantly—they will brown very quickly!
- When the marshmallows have toasted a warm brown color, have your adult helper remove them from the oven. **Caution:** the inside of the marshmallows will be extremely hot—do not eat or touch them until they have cooled!
- Examine the surface of your marshmallows. *Can you see the numbers that you painted on the marshmallows? Which numbers can you see? Which numbers are not visible? Are the numbers all the same color, or are some darker than others? Thinking about the various liquids, what do you see that might be similar about the liquids that made the lighter or darker numbers?*

EXTRA

Repeat this activity with other safe household products that you can paint onto your marshmallows. Try toothpaste, over-the-counter antacid remedies, or other types of fruit juice.

OBSERVATIONS AND RESULTS

In this activity, you examined how five different solutions affected the rate of marshmallow toasting. You should have observed that for some of the solutions the numbers written on the marshmallow appear darker than the rest of the marshmallow. In contrast, for other solutions, the numbers appear lighter than the rest of the marshmallow.

In the case of the darker numbers, the solution caused the marshmallow to toast faster in the area where you painted the marshmallow. This tells us that in this case when the number is darker the solution increased the rate of the Maillard reaction. In contrast, when the number on the marshmallow appears lighter than the rest of the marshmallow, we can determine that the solution decreased the rate of the Maillard reaction.

Did you notice any patterns related to which solutions caused the number to darken and which caused the numbers to lighten? It turns out that basic solutions (such as baking soda and milk) speed up (or catalyze) the rate of the Maillard reaction. In contrast, acidic solutions (such as lemon juice and vinegar) slow the rate of the Maillard reaction. Therefore you should have found that the numbers 2 and 4 looked darker than the surrounding marshmallow, whereas the numbers 3 and 5 appeared lighter than the rest of the marshmallow. Because water is relatively neutral, the number 1 may have appeared very faint, or not shown up at all on the marshmallow.

CLEANUP

Once the marshmallows have cooled completely, they can be handled and are safe to eat. Store them for later or to share with your helper.

Shattering Sugar

Make Movie-Ready Sugar Glass

LEARN ABOUT CHEMISTRY WHILE YOU MAKE SWEET "GLASS" THAT IS GOOD ENOUGH TO EAT—OR TO SHATTER!

PROJECT TIME

45 to 60 minutes

KEY CONCEPTS

Chemistry
Solubility
Temperature
Saturation

You're watching an action movie, and suddenly the hero dives through a glass window! Or a car window shatters as the hero navigates an exciting car chase! The glass looks so real, but believe it or not, movie sets rarely use real glass for those scenes. Can you guess what they use instead? If you guessed candy, you're right! Instead of using real glass, which is expensive, difficult to replace for multiple shots and dangerous, movie sets use sugar glass! This cheap, easy-to-make substitute looks like glass—and tastes like candy! In this activity you'll make your own beautiful version—a stained sugar glass window that looks good enough to eat!

BACKGROUND

You're probably pretty familiar with table sugar. You know what it tastes like—and probably what it looks and feels like too. You might even know that table sugar comes from sugar cane or sugar beets (that's right, 20 percent of the world's sugar comes from beets!). There are several different types of sugar, which are distinguished by their chemical structure. The chemical name for table sugar is sucrose. Sucrose is a relatively large sugar molecule, which can be broken down into smaller, simple sugars called glucose and fructose. For today's activity, when we say sugar we'll mean sucrose.

The table sugar crystals we can see and feel are called sugar granules. Table sugar granules are made up of millions of tiny sucrose molecules, all clustered together. One sugar cube has at least one billion trillion sucrose molecules!

If you've ever put a sugar cube or teaspoon of sugar into a cup of tea or lemonade, you know that sugar usually dissolves in water. When this happens you can't see the sugar in the liquid (even though you can taste it!).

When sugar is added to water (or water-based drinks), those big sugar granules break up into individual sucrose molecules that are too small to see. This happens because the individual sucrose molecules are attracted to the water molecules. Because sucrose and water molecules are attracted to each other, you can dissolve a lot of sugar in water. However, to get candy to be sweet, we need lots of sugar. This is tricky because just like a sponge can only hold so much water, water itself can only hold a limited amount of sugar. At a certain point there are too many sucrose molecules and not enough water molecules to pull them away from each other. When this happens we say the solution is saturated—it can't hold any more sugar. When the solution becomes saturated the sugar granules can't be pulled apart. When this happens we can see those undissolved sugar crystals in the water. So how do we get the leftover sugar to dissolve in the water, and make our candy extra sweet? We'll find out in this activity!

MATERIALS

- 1/8 teaspoon cream of tartar
- 1 cup (237 ml) of water
- 1 and 3/4 cups of granulated sugar
- 1/2 cup (118 ml) of light corn syrup
- Aluminum baking pan
- Non-stick cooking spray or parchment paper

- Saucepan
- Mixing bowl
- Cold water and a few ice cubes
- Oven or hot plate
- Food coloring
- Wooden spoon
- Metal fork (at least one)
- Candy thermometer (it's possible to do this activity without a thermometer, but we recommend using one if it's available.)
- Cooking mitt
- Clock or timer
- Piece of paper (or lab notebook) and pencil
- Adult helper to assist with handling materials at very high temperatures

PREPARATION

- Combine the cream of tartar, water, sugar and corn syrup in the sauce pan.
- Put several cups (about a liter, or 4 1/4 cups, depending on the size of the bowl) of cold water into your mixing bowl (until it is at least halfway full). Add a few ice cubes.
- Place the candy thermometer in the saucepan—make sure it's not touching the bottom of the pan.
- Spray the aluminum baking pan with cooking spray or cover with parchment paper. If using parchment paper, fold the edges so the paper fits into the pan.
- On your paper make a table with columns for temperature, time, appearance, feel, and other observations. Draw eight rows. Fill out each row of the temperature column as follows:

	Temperature	Time	Appearance	Feel	Other
Row 1	Room Temperature 1				
Row 2	230°F (110°C)				
Row 3	235°F (113°C)				
Row 4	245°F (118°C)				
Row 5	255°F (124°C)				
Row 6	275°F (135°C)				
Row 7	300°F (149°C)				
Row 8	Room Temperature 2				

PROCEDURE

- **Note:** Sugar syrup is extremely hot and sticky! The steps below should be done with the assistance and direct supervision of an adult helper.

- In Row 1 of your table, fill out the other four columns based on your observations. Use the wooden spoon (or your clean hands) to make note of the appearance and feel of the contents of the pot. Use the other observations column to note anything else that seems interesting. *How thick is the liquid in the pot? Would you compare it to water, or maple syrup? Is the solution clear? Can you see the separate ingredients, such as the sugar, or did it dissolve in the water? What do you predict will happen as you heat the solution?*

- Place your saucepan on the stove, and set to medium heat. Write the time that you started heating the liquid in the time column of Row 1. Be careful while heating the solution: to make a clear window you need to heat the sugar slowly, otherwise it will caramelize (turn brown). Use the wooden spoon to slowly stir the ingredients, and notice their consistency and appearance. Continue to stir the solution until it comes to a slow boil.

- When the temperature on the thermometer reaches 230°F (110°C), have your adult helper use the wooden spoon to carefully drop a small amount of the solution from the saucepan into your bowl of ice water. Record the time that the temperature reached 230°F (110°C) in Row 2 of your table.

- Wait until the sugar cools, and then use your hands to explore the drop of sugar solution in the water. Write your observations in Row 2 of your table. *What does it feel like? Does it dissolve in your fingers or keep a shape? Is it hard or soft? What does it look like? What does the solution in the pot look like? How has it changed since you started heating it?*

- Repeat these steps when the temperature on the thermometer reaches 235°F (113°C). Have your adult helper use the wooden spoon to carefully drop a small amount of the solution from the saucepan into your bowl of ice water. Record the time that the temperature reached 235°F (113°C) in Row 3 of your table.

- Wait until the sugar cools, and then use your hands to explore the drop of sugar solution in the water. Write your observations in Row 3 of your table. *How has the feel and appearance of the sugar changed? What does it feel like? Does it dissolve in your fingers or keep a shape? Is it hard or soft? Can you mold it to form different shapes? What happens if you try to remove it from the cold water? What does it look like? What does the solution in the pot look like? How has it changed since you started heating it?*

- Repeat these steps when the temperature reaches 245°F (118°C), 255°F (124°C), 275°F (135°C) and 300°F (149°C). If the temperature of the sugar syrup is not increasing, gradually increase the heat of the burner. As the solution heats up, allow sufficient time for the sugar to cool in the ice water before handling it. You can add ice to the bowl as needed to keep it cold.

- When the temperature is between 300°F (149°C) and 310°F (154°C), have your adult helper carefully pour the sugar mixture into the prepared foil pan. Spread the syrup evenly in the pan with a spoon.
- Drop food coloring across the surface of the hot syrup. Use your fork to swirl the colors. Have fun and be creative! But don't lick the fork—the syrup is very hot!
- Allow the syrup to cool completely.
- Carefully remove your sugar glass from the pan. Hold it up to a light source or window so you can see the colors clearly. Beautiful!
- Fill out the final row (Row 8) in your table. Note the difference in the appearance and feel of the sugar solution at each different temperature. *At what temperature did you notice the biggest changes in appearance? At what temperature did you notice the biggest changes in the feel of the syrup? Did the sugar heat up faster at certain points than it did at others?*

EXTRA

Different types of candy can be made by varying the temperature of the sugar and the rate that it heats and cools. You can try different candy making by removing the syrup at lower temperatures and observing what types of candy are made at each temperature stage. You can test higher temperatures but be careful, sugar burns at 350°F (177°C).

OBSERVATIONS AND RESULTS

In this activity, you observed the physical changes that sucrose goes through as it is dissolved in water and heated. When you began the activity at Room Temperature 1, the solution was saturated, and you were able to see the sugar crystals in the water. However, as you heated the sugar syrup, the sugar dissolved and the solution became clear so that you could no longer see the individual sugar crystals.

Why did heating the solution allow more sugar to dissolve in the water when it couldn't dissolve before? Heating up the solution causes the molecules to move around more and at greater speeds. As a result, the sugar and water molecules come in contact with each other more often, and with greater force. Imagine building three Lego cars, and putting the cars in a jar together. If you gently shake the jar, a few Lego pieces might fall off, but the cars will stay largely intact. However, if you shake the jar harder, the cars will bump into each other more often and with more force, and more pieces will break off. This is similar to what happens when you heat your sugar solution. Water molecules bump into the sugar molecules, causing them to break away from the larger sugar granule.

As the sugar dissolved in the solution, you should have observed other changes in the appearance and feel of the sugar syrup. At each recorded temperature point, the sugar in your solution was at a different stage. These stages are marked by changes in the physical appearance and form of the sugar. These changes take place as the water in the solution boils away and the sugar concentration of the solution increases. At 230°F (110°C), the sugar syrup was at the thread stage. At this stage your sugar syrup is approximately 80 percent sugar and 20 percent water, which is still too much water for the syrup to form a shape. When you dropped the syrup in cold water, it formed a very loose, liquid thread that probably didn't hold a shape very well. Even though you can't make candy at this stage, you can make a sugar syrup.

The next stage of sugar cooking is known as soft ball stage when the temperature of the syrup reaches 235°F (113°C). At this point, the sugar concentration in the syrup is approximately 85 percent and the name of this stage probably makes sense to you. The higher sugar concentration makes the syrup thicker, allowing it to hold a shape. When you dropped the sugar syrup into cold water, you were able to mold it into a soft, flexible ball. If you removed the ball from the water, you should have observed that it would lose its shape and flatten out. If you removed the sugar from the heat at this point, you could make fudge or pralines!

The next time you tested the syrup, at 245°F (118°C), it had reached the firm ball stage. Can you guess why this stage has that name? Hopefully you can because when you dropped the syrup in cold water at this stage, you could form it into a firm ball that held its shape when removed from the water. At this point, the syrup is approximately 87 percent sugar, and just that small difference in the sugar concentration results in a dramatic change in the firmness of the syrup.

The next stage is called hard ball stage, reached when the syrup heats to 255°F (124°C). At this point the sugar concentration is 92 percent, and as you dropped the syrup in cold water you might have observed that it formed thick, ropelike threads as it dripped from the spoon. Because it has less moisture at this stage, once the syrup reached the cold water it should have formed a hard ball that was more difficult to mold into different shapes. Marshmallows and gummies are made from sugar at the hard ball stage.

When the syrup reached 275°F (135°C), you might have noticed the syrup in the saucepan had smaller, thicker bubbles. This stage is known as soft crack stage, and the sugar concentration is very high, approximately 95 percent. As a result, when you dropped the syrup into cold water it should have formed threads that could be bent slightly before they would break. If you wanted to make taffy, you would take the sugar off the heat at this point.

The final stage in this activity is hard crack, reached when your syrup heated to 300°F (149°C). At this temperature, almost all of the water has boiled off, so the solution is 99 percent sugar. When you dropped it in the cold water, it should have formed hard, brittle threads that would break easily. This is the stage used to make toffee and hard candies like lollipops—or sugar glass like they use in the movies!

You might be wondering why we added the other ingredients: corn syrup and cream of tartar. Adding these ingredients helps to create the clear, glasslike structure we want in this activity. The corn syrup helps prevent the sugar molecules from joining back together and forming large granules (or crystals) as the syrup cools. The cream of tartar helps break the sucrose molecules up even father, into the simple sugars glucose and fructose. And the food dye makes the whole thing a work of art!

CLEANUP

Wipe down surfaces and clean containers you have used. Replace materials where you found them in your home. Discard any of your creations that you are not going to eat.

Candy Lightning

FIND CANDY THAT PACKS A FLASH—AND LEARN WHAT CAUSES THIS BRIGHT REACTION.

Did you know that instances of miniature lightning can occur in your mouth when you bite on a hard candy? Would it not be fascinating if you could observe these light flashes? Life Savers Wint-O-Green Mints might just be the tool we need to observe this science! Crush them, observe and detect what makes the flashes visible. How many sparks can you see?

PROJECT TIME

30 to 40 minutes

KEY CONCEPTS

Chemistry
Light
Static electricity
Fluorescence

BACKGROUND

Lightning is a massive release of static electricity. It starts with icy raindrops rubbing against the air. This rubbing gives rise to static electricity, and sometimes the electricity can jump across an air gap. This is called a discharge, or spark. Within clouds or between clouds and Earth the discharge is huge; it heats the air to the extent that it glows with white light. That glowing white light is the lightning you see.

Miniature buildups of static electricity occur often. Maybe you remember your hair standing up after brushing it or clothes clinging together when they come out of the dryer. In both examples, rubbing created static electricity. Crushing particular crystalline structures like hard sugar or salt also gives rise to static electricity.

Whenever there is static electricity, a release might happen. You might have felt or seen a discharge when touching a metal doorknob after walking across a carpeted floor. Although releases occur often, we hardly ever see a light flash. The miniature releases excite the air around it, which then typically releases ultraviolet (UV) light, a type of light humans cannot see. It is only when there is a fluorescent substance around, which can absorb the UV light and reemit it as visible light, that we see a tiny light flash. And this is exactly what happens in lightning—and this activity as well.

MATERIALS

- Life Savers Mints: Wint-O-Green
- Sugar Free Life Savers Mints: Wint-O-Green
- Life Savers Mints: Spear-O-Mint and/or Pep-O-Mint
- Other hard candy (optional)
- Pliers
- Dark room

PREPARATION

- Bring all of your materials to the room you will be using for this activity, then prepare to turn off the light.
- **Note:** The electrical discharges created in this activity are tiny and harmless. Although it is the same phenomenon, lightning in storms is created by huge discharges and is dangerous.

PROCEDURE

- Place one Life Savers mint between a pair of pliers and turn off the light. Wait a minute or two to let your eyes adjust to the dark, then watch the candy as you crush it. *Did you see anything?*

- Repeat with all other types of mints. *Did you see anything happening as you crushed some types of Life Savers whereas it did not happen when you crushed other types?*

- To confirm your findings, repeat the activity a few times. *Does the same phenomenon happen each time with the same type(s) of Life Savers Mint and not with others?*

- You probably detected it for only one of the types of Life Savers you crushed. *Can you find which combination of ingredients is responsible for the phenomenon?* You will need some detective skills to find out! *Where can you find the ingredients of each type of candy? In what way(s) are the ingredients of the mint that shows the phenomenon different?*

- Now, we will investigate what exactly creates the flash. Take another Life Savers Mint, the type that exhibited the phenomenon, and start exploring. We know the flash happens when we crush the candy with pliers. *Does it also happen when you rub two broken pieces together or when you press just enough on the pliers to put the candy under pressure without breaking it?*

SCIENCE FAIR IDEA

Investigate other candies containing hard crystalline sugar and the special ingredient. *Can you create a flash with those as well?*

EXTRA

If you are wondering if it happens when you chew the candy, try it out. Chew this type of Life Saver in a dark room in front of a mirror. Open your mouth wide between chews. *Can you see a flash?*

EXTRA

Try to create miniature lightning. Vigorously rub a piece of paper with wool for about a minute. Put a piece of metal (such as aluminum foil) in the center of the paper. Remove the paper and bring your finger close to the metal. *Can you create a tiny spark?*

OBSERVATIONS AND RESULTS

Did you see a light flash when crushing the sugar-containing Life Savers Wint-O-Green Mints but not when crushing the other types of Life savers Mints? This is expected. You need hard crystalline sugar as well as the artificial flavoring substance in the Wint-O-Green type to create the flash.

Crushing hard crystalline sugar creates static electricity, which can lead to a mini discharge, a kind of miniature lightening. You rarely see a light flash because the discharge is usually accompanied by ultraviolet (UV) light, a type of light humans cannot see. Note that many animals, like insects and birds, would observe this UV light flash because their eyes are sensitive to UV—but human eyes are not.

Life Savers Wint-O-Green Mints (and maybe a few other candies) contain a fluorescent substance. Fluorescent substances absorb UV light and release it again as light humans can see. When crushing Life Savers Wint-O-Green Mints, the UV light created from crushing the hard crystalline sugar gets absorbed by the fluorescent substance and released again as visible light humans can observe. Voilà—there is your light flash!

Although lots of hard crystalline structures emit UV light when crushed, the crystals in the sugar substitute do not. Therefore, no UV flash and thus no visible flash was seen when crushing Sugar Free Life Savers Wint-O-Green Mints.

CLEANUP

Wash any sticky parts of the pliers with soapy water and then dry thoroughly. Clean up any candy fragments in your work area.

Make Your Own Gelatin Pearls

TRY YOUR HAND AT MOLECULAR GASTRONOMY WITH THIS COLORFUL KITCHEN CHEMISTRY ACTIVITY!

Do you enjoy getting creative in the kitchen? If so, this activity is for you! Molecular cuisine—taking tools, ingredients and methods typically used in science and using them in cooking—might sound fancy and complicated, but some techniques are easy to replicate! Get your hands wet, fire up your creativity and see how rewarding it can be!

PROJECT TIME

45 to 50 minutes

KEY CONCEPTS

Chemistry
Molecular gastronomy
Surface tension
Spherification

BACKGROUND

Molecular gastronomy—the science of culinary phenomena—brings scientific procedures, ingredients and instruments into the kitchen. Part of this field studies the influence food preparation techniques have on texture and flavor. Spherification of juices, changing a liquid into semisolid pearls, is one example. The most common method uses two chemicals, such as sodium alginate and calcium chloride: One chemical is dissolved in the liquid and another in a water bath. When drops of the treated liquid fall into the bath, they form small spheres. The outside of these spheres solidifies as a result of a chemical reaction between the two additives. There you have it—little drops of liquid confined in a thin, solid membrane!

One can also make gel-like spheres using more common ingredients: gelatin and oil. Gelatin is made of the protein collagen, a large fibrous molecule abundant in many animals, including humans. Collagen makes skin, bones and tendons strong and somewhat elastic. This large protein is partially broken down and further treated to form gelatin. When gelatin is prepared, its proteins unravel as a response to added heat. These long strings interlace when cooled, however, and create a three-dimensional structure that can trap a lot of liquid. If prepared with water, the resulting gel is transparent, colorless and flavorless. In the kitchen, it is often used for its unique texture.

MATERIALS

- Unflavored gelatin powder, two 2 1/2-ounce packages
- Water or juice, 1/2 cup (118 ml)
- Spoon
- Food coloring (red, blue, or green work well)
- Vegetable oil, 4 cups (0.9 l)
- Water
- Four tall containers
- Freezer
- Ice cubes
- Saucepan or microwavable cup
- Stove or microwave
- Dropper or syringe
- Slotted spoon
- Plate
- An adult helper
- Work area that can get wet
- Red cabbage (optional)

PREPARATION

- Pour about 2 cups (473 ml) of vegetable oil into each of two tall containers.
- Place one container in the freezer for at least 30 minutes.
- Fill the other two containers with about 2 cups (473 ml) of water each.
- Add ice cubes to one container to cool the water.
- To prepare the gelatin, pour 1/2 cup (118 ml) of liquid (water or juice) into the cup or saucepan. Add two packages of gelatin powder and stir until all the powder is dissolved.
- Add food coloring so you can easily see the pearls in the water or oil. Have an adult helper carefully heat the liquid on the stove or in the microwave until it is very hot while stirring occasionally. Let the gelatin cool for 5 to 10 minutes.

PROCEDURE

- Have the dropper or syringe, the four containers—two filled with water and two filled with oil—and the warm gelatin nearby. In a minute you will put a few drops of warm gelatin into each container. *What do you think will happen?*
- Use the dropper/syringe to put a drop of warm gelatin into each of the containers and watch what happens. Repeat this until you have added three drops to each container. *Do you see little pearls form in some liquids and not in others? Does something similar happen each time? Why do you think this happens?*
- Let the drops sit for a minute, then see if you can scoop the pearls out with a slotted spoon. *Which liquid allows you to form gelatin pearls? Why do you think this is the case?*

- Rinse the pearls with water and try tasting one. *How do your gelatin pearls taste? What is the texture like?*
- Use the leftover warm gelatin to make more gelatin pearls. Try to make different size pearls or gelatin spaghetti. Recool the oil that was in the freezer in between batches if needed.

EXTRA

Watch your gelatin pearls shrink as they dry out. Put them back into water to see them swell again. *Why do you think this would happen?*

EXTRA

Prepare red cabbage juice by boiling the cabbage in plenty of water. After the water has turned purple, use it to make red cabbage juice pearls. Discover how these pearls are special by placing them into vinegar and/or a baking soda solution. *Can you also explain your observations?*

EXTRA

Look up other spherification methods using sodium alginate and calcium chloride. Order these chemicals or a spherification kit and make juice balls or yogurt ravioli.

OBSERVATIONS AND RESULTS

Did droplets of warm gelatin only form nice pearls in the cold oil whereas dropping them into water or room temperature oil failed to form pearls?

To prepare the gelatin, you dispersed gelatin particles homogeneously in the liquid you used for this activity (water or juice). These particles unravel into long strings when heated. If they are concentrated enough, a gel forms as the gelatin interlaces and creates a three-dimensional structure that traps liquid as it cools. When warm gelatin is dropped in water, the particles spread out over all the water and their concentration becomes too low to solidify. When warm gelatin is dropped in oil however, the gelatin stays together in little spheres. This happens because water and oil do not mix. Gelatin particles shy away from oil so they clump together in the shape that gives the least possible interaction with oil: a sphere.

In addition to the concentration, the temperature of the solution matters. Gelatin needs to cool below a specific temperature to solidify. Cold oil provides the necessary cooling, but room temperature oil does not. As a result, you can scoop out nice solid gelatin pearls from the cold oil bath—but neither from the room temperature oil bath nor the water baths.

Can you improve the taste of your gelatin pearls? What oil and what liquid would make a great combination? Can you also change the texture?

CLEANUP

Pour the cooking oil into a container and bring it to your local recycling center or dispose of it in your garbage. (Never pour cooking oil down the sink drain! It clings to the pipes and solidifies in time, leading to pipe blockages.) Wash all equipment with water and soap.

Make Mouthwatering Candy Melt

WHAT ARE THE PROPERTIES OF YOUR FAVORITE CANDIES? USE CHEMISTRY TO TEST DIFFERENT TYPES—AND SEE WHAT IT IS THAT MAKES YOUR MOUTH WATER!

What is your favorite type of candy? Have you ever wondered why you like this candy so much? Is it its taste, how it feels in your mouth—or both? Cooks and food scientists study how substances dissolve or melt to create a unique and pleasant sensation in the mouth and optimize the release of flavor. Would you like to know how your favorite candy works? In this science activity you will study two types of candy and discover what makes them so enjoyable.

PROJECT TIME

35 to 45 minutes

KEY CONCEPTS

Chemistry
Food science
Solubility
Gelatin

BACKGROUND

The digestion of food is one big chemical process. It starts in our mouths where saliva—a watery liquid—surrounds the food. Saliva helps us to chew and swallow food and contains a small quantity of enzymes, or chemicals that help break down our food. How a food interacts with saliva will influence how it tastes as well as how it feels in the mouth. Will it dissolve, will it melt or will it just be chomped into pieces by our teeth?

Whether a solid dissolves when surrounded by water depends on how strong the internal bonds are and how eager it is to bond with water. For example, to dissolve sugar in water, bonds between sugar molecules (the tiniest sugar particles) and between water molecules (the tiniest water particles) must be broken. This requires energy. Luckily the creation of new bonds between sugar and water releases enough energy to break those bonds so sugar can dissolve in water. If the released energy is less than the energy required, however, the solid will not dissolve.

Temperature plays a role as well. Adding heat means adding energy—and for most solids the hotter the water, the easier and faster it is to dissolve the solid. Heated molecules move around more so the dissolved substance will disperse more quickly than in unheated water.

Most candy contains sugar and/or corn syrup as a sweetener with a few other ingredients to add flavor and texture. Hard rock candies, such as fruit drops, are mainly crystallized sugar that dissolve when surrounded by saliva. For soft candy, such as gummy bears, gelatin is often used to create their particular texture. Gelatin is derived from collagen, which is a structural protein found in all animals. Gelatin protein gets tangled when cooked, creating mesh pockets that trap water and other ingredients. After gelatin cools the proteins remained tangled and provide structure that holds many desserts or candies together. For most types of gelatin, their melting point is below body temperature, so they melt in the mouth and provide unique sensations.

MATERIALS

- Four glasses
- Cold water
- Hot water
- Pen
- Sticky notes
- Gummy bears
- Hard candy, such as fruit drops
- Two timers or a clock
- Spoons
- Two plates
- Work space that can tolerate spills
- Ruler (optional)

PREPARATION

- Protect your work space for splashes of water.
- Fill two glasses each with 1/3 cup (78 ml) of cold water.
- Label two sticky notes with "Cold Water" and two more with "Hot Water."

PROCEDURE

- First, we will observe what candy does when placed in cold water. Place two gummy bears in the first glass of water.
- Then remove the wrappers of two hard candies, and place them in the second glass.
- Identify this row of glasses with a sticky note reading "Cold Water."
- Set the first timer to 30 minutes or use the clock to track 30 minutes.
- *What do you think will happen to the candy if left untouched for 30 minutes? Why do you think this will happen, and will the water change?*
- While you are waiting fill the two additional glasses with 1/3 cup (79 ml) of hot water. (Be careful not to get any hot water on your hands.)
- Carefully place two gummy bears in the first glass.
- Then remove the wrappers of two hard candies and place them in the second glass.
- Identify this row of glasses with a sticky note reading "Hot Water." Set the second timer to 30 minutes or use the clock to track 30 minutes.

- *What do you think will happen to this candy if left untouched for 30 minutes? Do you expect it to be different from the candy in the cold water? Why would this be the case? Do you expect the water to change?*
- Label one plate "Cold Water" and the other "Hot Water."
- After the first 30 minutes are over, it is time to observe the candy submerged in cold water. Use a spoon to carefully pick the candies out of the two glasses in the Cold Water row and place them on the plate labeled "Cold Water." Set aside.
- For each of these cold water glasses, take a clean spoon to stir the water, scoop up a sample and taste it. *Which one is sweetest?*
- Take a fresh gummy bear and a hard candy out of the bag (removing the wrapper if needed), and place the candy next to the candies that have been soaked for 30 minutes. *What differences do you observe? Is what you observe consistent with what you expected? If it is different, why do you think this is the case?*
- *Looking at the change in the size of the candies and the taste of the water, what do you think happened to each type of candy when submerged in water?*
- After the second timer goes off and the Hot Water row candies have been submerged for 30 minutes, repeat the observation procedure for these candies. Remember to use the plate labeled "Hot Water" this time.
- Now that we have seen what cold and hot water does to these types of candy, let us look closer at the influence of temperature. *How is the candy soaked in hot water different from the same type of candy soaked in cold water? Why do you think this has occurred?*

- Pop a fresh candy from the bag into your mouth. Notice how your mouth is a watery environment that is at body temperature. *How do you think your previous observations relate to the feel and taste of the candies in your mouth?*

SCIENCE FAIR IDEA

Study the influence of heat further by filling a glass with 1/3 cup (79 ml) of ice-cold water and another glass with 1/3 cup (79 ml) of hot water. If possible, place the hot water glass in a hot water bath, so the water does not cool off as fast. *Put a brightly colored hard candy in each glass and watch how the color spreads in each glass. Which one spreads faster? Why do you think this happens?*

SCIENCE FAIR IDEA

Use a ruler to measure the lengths of the unsoaked and soaked candies, then express the change as a percentage. *Can you estimate the change in volume as well?* You can also compare the sizes of the candies soaked in hot water with those soaked in cold water.

SCIENCE FAIR IDEA

Study the effect of additives such as salt or sugar in the water. *Would a gummy bear soak up more, less, or an equal amount of water if soaked in a salty or sugary solution compared with tap or distilled water? Why do you think this happens?*

EXTRA

Add chopped pieces of fresh tropical fruits such as pineapple, kiwi, or papaya to the water. *Would the gummy bears dissolve differently when these fruits are added to the solution? What about if we add apple, lemon, or orange slices?* (Hint—think of gelatin desserts and why certain fresh foods cannot be added to them.)

OBSERVATIONS AND RESULTS

Did you see that the hard candies dissolved in cold water whereas the gummy bears puffed up? Did hot water dissolve the hard candies faster whereas the gummy bears started to melt in the hot water?

Hard candy is often made primarily of sugar, corn syrup, and other substances that dissolve easily in water. Like the saliva in your mouth, the water in the glass allowed them to dissolve. Adding heat made this process easier and faster. As the candy dissolved layer by layer, it got smaller and smaller, and more and more sugar, syrup, and flavor chemicals dispersed in the water. That is why the water tasted sweet, like the candies.

Gummy bears, on the other hand, contain gelatin, which gives it its typical soft texture. The gelatin creates a mesh of entangled long molecules that hold the gummies together. The structure has pockets that can absorb water via a process called osmosis. That is why the gummies swell up when soaked in cold water. The water stays relatively tasteless. It was different when you submerged the gummies in hot water. Because the temperature of the hot water was above the gelatin's melting temperature, the outer layer of the gummies melted away. The gummy bears got smaller and smaller, and you could see tiny pieces of gummy bear floating in the water. These pieces make the water taste sweet, like the gummy bears.

Hot water molecules move around more so you will see that the color disperses more in hot than cold water. Fresh papaya, pineapple, and kiwi contain enzymes that break down gelatin. The addition of these and a few other enzyme-containing fruits can make your gummy bears break down.

CLEANUP

Throw away soaked candy and wash all equipment with soapy water.

Making a Sugar Thermometer

HOW CAN YOU MEASURE TEMPERATURE WITHOUT A THERMOMETER? FIND OUT WITH THIS SWEET SCIENCE ACTIVITY!

PROJECT TIME

45 to 60 minutes

KEY CONCEPTS

Chemistry
Temperature
Crystallization
Calibration

Bakers (and those who help bakers!) know that at some point in every baking recipe the instructions will tell you to preheat your oven to a certain temperature. But if you've ever tried to bake cookies and they come out flat or take too long, it's possible that your oven is to blame. You might set your oven to 350°F (177°C), but how do you know that the inside of your oven actually reaches that temperature? You could use a thermometer—or you could use sugar! Because we know that table sugar (known as sucrose) melts at a very specific temperature, we can test the accuracy of an oven by experimenting to see if sugar melts when the oven is set to the correct melting point. In this activity you will use table sugar to make a homemade oven thermometer and put your oven to the test!

BACKGROUND

If someone asked you to name foods containing sugar, you could probably name quite a few. But would beets be one of the foods you listed? If not, then you might be surprised to learn that about 50 percent of the sugar in our foods comes from sugar beets! The vast majority of the sugar we eat in processed food comes from either sugar beets or sugarcane. But a lot of work goes into extracting and purifying the sugar from these plants. In fact, processing sugar was so expensive and difficult that even just a few hundred years ago refined sugar was considered a fine and very expensive spice, and only the wealthiest people could afford to eat it.

To extract sugar from sugarcane or sugar beet plants, the plants are soaked in very hot water and squeezed through a press to extract a sugar syrup. When this syrup cools, it forms the sugar crystals (sucrose) that we buy at the grocery store.

Because the process of making sugar involves heating and cooling it, sugar manufacturers have learned that the melting point of sucrose (table sugar) is 367°F (186°C). Knowing this information, we can use sugar to test the accuracy of your oven's temperature. In this activity, you'll use sugar to test whether your oven is as hot as it says it is!

MATERIALS

- Aluminum foil, cut into two 6- by 6-inch (15- by 15-cm) squares
- White sugar (at least 3 tablespoons)
- A baking tray
- Oven mitts
- Three plates
- An oven
- An adult to help

PREPARATION

- In this activity, you will be working with materials heated to very high temperatures. This should only be done with the help and under the direct supervision of an adult.
- Preheat your oven to 350°F (177°C).

- Fold the outer about 1/2-inch (1.3 cm) edge of each aluminum foil square up to make a miniature dish that is about 5- by 5- inches (12.7- by 12.7-cm) square and about 1/2 inch (1.3 cm) tall.
- Add a tablespoon of sugar to each sample dish.

PROCEDURE

- Add a tablespoon of sugar to each aluminum foil sample dish. Notice the consistency of the sugar. *What words would you use to describe what the sugar looks and feels like?*
- Once the oven temperature has reached 350°F (177°C), put the first foil sample dish into the oven.
- Set a timer for 20 minutes and wait. While you're waiting, put one tablespoon of sugar on one of your plates. As you look at this sugar, make predictions about what you think will happen to the sugar being heated in the oven. *Will heating the sugar affect how it looks? What about how it smells, feels, or tastes? Can you think of other things that look or feel different when they're heated?*
- After 20 minutes, remove the sugar from the oven and transfer it to a different plate. Remember, the sugar is hot! Observe the sugar in sample dish number one. *Does it look any different after being in the hot oven? Is this what you predicted or did you expect something different?*
- Set your oven to 375°F (191°C). When it reaches that temperature, put the second sample dish into the oven, and set a timer for 20 minutes. *What do you think will happen to the sugar in the second sample dish? Do you think it will be the same as the first experiment or something different?*
- After 20 minutes remove the second sample dish and transfer it to your third plate. Remember, the sugar is hot!

- Carefully set the plates side by side. Compare the three samples. *What differences do you see between the two baked samples? Are they different from each other? If so, what is different about them? What do you think accounts for the differences between the samples? What differences do you notice between the unbaked sample and the other two samples? Is the uncooked sample similar to either of the cooked ones? Why might this be?*
- Once the samples have cooled down, taste them! Notice if there are any differences in how each sample tastes.
- **Note:** If you did not see any changes to the appearance of the sugar in either test, try repeating the experiment with the oven temperature set to 400°F (204°C).

EXTRA

Try repeating this activity using other temperatures. For each test, try getting a little closer to sugar's melting point (367°F [186°C]). If your sugar doesn't melt when your oven is set to 365°F (185°C) but does melt when it's set to 370°F (188°C), you'll know that you have a very accurate oven!

EXTRA

Repeat the test using oven-safe bowls of water, to test how accurate your oven is at lower temperatures. The boiling point of water is 212°F (100°C), so you could try repeating the test with your oven set at 200°F (93°C) and then 225°F (107°C). *Does the water boil at one temperature but not the other?* If so, you have a very accurate oven! Remember, only do these tests with the help and under the direct supervision of an adult!

OBSERVATIONS AND RESULTS

In this activity, you baked two different samples of sugar at two different temperatures. Knowing that sucrose (table sugar) melts at exactly 367°F (186°C) allows us to use sugar as a very simple thermometer. If your oven is accurate, baking the sugar at 350°F (177°C) should not have significantly changed the appearance of the sugar, because 350°F (177°C) is 17 degrees below sugar's melting point. In the second part of the activity, however, baking the sugar at 375°F (191°C) (which is seven degrees higher than its melting point) should have caused it to melt, making the sugar look almost like glass. If this is what you saw during your experiment, you know you have a very accurate oven!

Don't worry if the sugar didn't melt when you tested it at 375°F (191°C) degrees, however. This means that you've found a very interesting result! Your oven temperature isn't correctly calibrated. (Your oven thinks it's heating to 375°F [191°C], but it's actually not getting that hot!). Hopefully you repeated the test with the oven set to 400°F (204°C). If your sugar still didn't melt at that setting, you know that your oven's temperature is off by over 30 degrees! Make sure to keep this in mind next time you're baking cookies!

CLEANUP

Wash all equipment with water and soap.

Maple Syrup Science

Cooking Up Some Candy

HAVE YOU EVER COOKED UP HOMEMADE CANDY? MAYBE FROM CHOCOLATE—OR TABLE SUGAR?

PROJECT TIME

45 to 60 minutes

KEY CONCEPTS

Chemistry
Cooking
Food science
Solutions

Maple syrup is not only deliciously gooey and great on breakfast foods such as pancakes and waffles, it can also amazingly turn into maple candies with a range of textures. For example, it can be made into sticky maple taffy or hard, molded maple sugar candy. Early spring is the time of year when maple producers in North America start to tap their trees and collect the sap to turn it into maple syrup, so it's a perfect time to explore this tasty, fascinating food. In this science activity, you will investigate how the temperature of heated-up maple syrup affects what types of candies can be made from it.

BACKGROUND

Chemically speaking, maple syrup is a concentrated solution of sugar in water, with many minor flavoring compounds. It is made by collecting sap from certain maple trees and boiling it. The sap contains a lot of sugar, primarily in the form of sucrose, although there are also small amounts of glucose and fructose. The sap is gently boiled until it is the right consistency—not too watery (or the maple syrup could easily spoil) and not too thick (or the sugar in the maple syrup could harden and crystallize).

How does boiling the sap change its consistency? When a sugary solution is heated, some of the water evaporates, and this makes the sugar more concentrated in the solution and the overall product thicker. As the heated maple syrup cools, sugar molecules (the smallest particles of the sugar) can form crystals. The amount of crystals that form depends on how concentrated the sugar is.

This same principle is used to turn maple syrup into different types of sugary candies. To make maple candies, maple syrup is heated more, and this makes the sugar in the candies even more concentrated than it is in the maple syrup!

MATERIALS

- Baking pan or hard candy molds
- Cooking oil if using candy molds. If you are using a baking pan, you can use wax paper instead of cooking oil.
- 1/2 cup (237 ml) of pure maple syrup (make sure it is not imitation syrup!)
- Stove (use adult assistance when using the stove and handling hot objects)
- Saucepan
- Large spoon for stirring
- Candy thermometer or digital thermometer that can withstand temperatures of at least 310°F (154°C)
- A metal measuring teaspoon

PREPARATION

- Place your baking pan or hard candy molds on the counter next to the stove.

- If you are using a baking pan, cover it with a sheet of wax paper or lightly grease it with cooking oil. If you are using hard candy molds, lightly grease the molds with cooking oil.
- Make sure an adult helps and supervises while using the stove and heating the maple syrup. Also be sure to only use a metal measuring teaspoon, because the heated maple syrup can melt plastics!

PROCEDURE

- Heat 1/2 cup (237 ml) of maple syrup over medium heat in the saucepan, uncovered, using a large spoon to stir it occasionally at first. Use a thermometer to carefully measure the temperature of the maple syrup as you begin to heat it up. *How quickly does the temperature of the maple syrup increase?*
- Bring the maple syrup to a boil and allow it to cook, uncovered, while you stir frequently to make sure it does not burn. Carefully use the thermometer to measure the temperature of the maple syrup as you continue to heat it. *How quickly does the temperature increase now?*
- When the temperature of the maple syrup reaches 230 to 235°F (110 to 113°C), carefully use a metal measuring teaspoon to take an approximately 1 teaspoon (5 ml) sample of the heated maple syrup. Quickly pour the teaspoon of heated maple syrup onto the baking pan or candy mold. Do not touch the candies until they have cooled! *How easy is it to scoop and pour the heated maple syrup? What is its consistency like?*
- Similarly, take four more samples as the maple syrup heats, specifically one sample each at 245 to 250°F (118 to 121°C), 255 to 260°F (124 to 127°C), 270 to 280°F (132 to 138°C), and 300 to 310°F (149 to 154°C). Keep track of which candy came from which temperature sample. *How easy is it to scoop and pour each of the different samples? How are they similar and different from each other? What do you notice about the maple syrup in the saucepan as the syrup is heated up?*

- When you have made your five samples, let them all cool to room temperature and then investigate them. *What is the color of each sample? Is it transparent? Does it feel sticky, smooth and/or rough? Does it seem caramelized? Is it flexible, or does it break when bent?*
- When you are done making your visual observations, try tasting—or having a volunteer taste—each sample. *How does each one taste? Which is the sweetest? Which has the best texture? Which one do you like the most overall?*
- *Overall, how does it seem that heating the maple syrup affects the type of candy it can turn into? How are these candies similar to, and different from, maple syrup-based candies that you can buy?*

SCIENCE FAIR IDEA

Repeat this activity using imitation maple syrup instead of real, pure maple syrup. *Does imitation maple syrup make candies the same way that pure maple syrup does? If it makes different candies, how are they different?*

SCIENCE FAIR IDEA

You could try this activity again, but this time make several samples (at least three) at each temperature you test, and invite friends and family for taste testing. *Which samples do other people prefer? Is it always the same, or do some people prefer different aspects?*

EXTRA

Maple syrup is mostly made of the sugar sucrose and water. There are many different types of sugars, and you could design a way to make solutions of these other sugars and similarly test them. *Do they make candies similar to the ones made with heated maple syrup?*

EXTRA

Another way to make different candy samples is by doing the "cold water candy test." Try making maple candy samples using maple syrup heated to different stages according to this test. *How do the candy samples compare to each other? How are they similar to, or different from, the candy samples made in the first activity?*

OBSERVATIONS AND RESULTS

Did the hottest temperatures make the darkest, stiffest, sweetest candies, while the lowest temperatures made honey-colored, transparent, sticky candies?

Just as maple syrup is made by boiling and concentrating sap from trees, maple syrup-based candies are made by heating up maple syrup to concentrate it further. The hotter and longer the syrup is heated, the more concentrated the sugar (and other ingredients) in it becomes as the water evaporates away. This is why you should have seen that the candies made from maple syrup that was heated the longest (to around 270 to 310°F [132 to 154°C]) was the darkest (a light or dark brown color), non-transparent, rough in texture, stiff when you tried to bend them (breaking cleanly), but quite sweet! Meanwhile, the candies made from maple syrup that was not heated as long (to around 230 to 250°F [110 to 121°C]) should have been the most like maple syrup—they should have been honey-colored, transparent, sticky but smooth in appearance, and very flexible, but not super sweet.

When you think about your results, you compare about the difference between the texture of caramels and lollipops—the caramel is softer and chewier, while the lollipop is hard and cracks when it is bitten. The syrup used to make caramels is cooked until it reaches 240 to 250°F (116 to 121°C), at which point is has an 87 percent concentration of sugar and enters the "firm-ball stage." The syrup used to make lollipops, however, is cooked until it has a 99 percent concentration of sugar (at a temperature of 300 to 310°F [149 to 154°C]), which makes drops of the syrup turn into hard, brittle, easy-to-break threads when placed in cold water.

CLEANUP

To clean the saucepan, spoons, stove, or other kitchen items that may have gotten hot maple syrup on them, soak them with hot, soapy water and then scrub the remaining maple syrup off.

Scrumptious Science
Making Ice Cream in a Bag

HAVE YOU EVER MADE HOMEMADE ICE CREAM? IT CAN BE A LOT OF FUN, AND YOU END UP WITH A TASTY FROZEN TREAT!

A lot of interesting chemistry is actually needed to make ice cream. For example, think about how you start out with refrigerated (or even room-temperature) ingredients and then need to cool them down to turn them into ice cream. How do the ingredients change during this process? How important do you think it is that they get cooled to a certain temperature? In this science activity, you'll make your own ice cream (in a bag!) and explore the best way to chill the ingredients to make them become a delicious reward!

PROJECT TIME

40 to 50 minutes

KEY CONCEPTS

Food science
Chemistry
Freezing
Phases of matter
Solutions

BACKGROUND

To make ice cream, the ingredients—typically milk (or half and half), sugar, and vanilla extract—need to be cooled down. One way to do this is by using salt. If you live in a cold climate, you may have seen trucks spreading salt and sand on the streets in the wintertime to prevent roads from getting slick after snow or ice. Why is this? The salt lowers the temperature at which water freezes, so with salt, ice will melt even when the temperature is below the normal freezing point of water.

Technically, the temperature that the salt lowers is called the freezing point. When a freezing point is lowered, such as by adding salt to water, the process is called freezing-point depression. As we'll see in this activity, freezing-point depression is not unique to solutions made of water and salt; it also happens with other solutions. (A solution is made when a substance, such as salt, is dissolved and becomes a solute. The medium into which it is dissolved is a solvent—typically a liquid, like water.)

MATERIALS

- Measuring spoons
- Measuring cup
- Sugar
- Half and half (Alternatively, milk or heavy whipping cream may be used.)
- Vanilla extract
- Salt (Different types of salts, such as table salt or rock salt, should work but may give slightly different results.)
- Two small, sealable bags such as pint-size or sandwich-size Ziplocs
- Two gallon-size (3.8 l) sealable bags
- 8 cups (1.8 l) of ice cubes
- Oven mitts or a small towel
- Timer or clock

PREPARATION

- In each small bag, place 1 tablespoon of sugar, 1/2 cup (118 ml) of half and half (or milk or heavy whipping cream) and 1/4 teaspoon (1.2 ml) of vanilla extract. Seal up each bag after adding the ingredients. Keep the bags in the refrigerator until you are ready to continue on to the procedure.

PROCEDURE

- Add 4 cups of ice cubes (a little more than 1 liter) to one of the large, gallon-size bags. Then add 1/2 cup of salt to the bag. *What do you think the salt will do?*

- Put one of the small bags you prepared into the large bag with the ice cubes. Be sure both bags are sealed shut.

- Put on oven mitts or wrap the bag in a small towel and then shake the bag for five minutes. Feel the smaller bag every couple of minutes while you shake it and take a peek at it. *What happens to the ingredients over time? When five minutes are up, how do the ingredients look? What about the ice cubes—how do they change over time and how do they look by the end?*

- Now add 4 cups (about 1 liter) of ice cubes to the other large, gallon-sized bag, but this time do not add any salt to it. *What do you think will happen without using salt?*

- Put the other small bag you prepared into this large bag. Be sure both bags are sealed.

- Put on oven mitts or wrap the bag in a small towel and then shake the bag for five minutes, as you did before. Again, feel the smaller bag every couple of minutes while you shake it, and take a peek at it. *What happens to the ingredients over time now? When five minutes are up, how do they look now compared with last time? What about the ice cubes—did they change in the same way?*

- You can also compare how cold the different ice cube bags feel. *Does one feel much colder than the other?*

- If you successfully made some ice cream, you can enjoy it now as a tasty reward for your chemistry challenge! If the ingredients in one of your bags did not become ice cream, check out the Extras for tips on turning them into ice cream.

EXTRA

If one of your bags did not make ice cream, try putting it back inside the large bag that had ice cubes and salt and then shake it for five minutes. *Did it turn into ice cream? Why do you think you got the results that you did?*

EXTRA

Instead of half and half, you could use milk or heavy whipping cream to make ice cream in this activity. *How does making ice cream with half and half compare with using milk or heavy whipping cream?*

EXTRA

You can make ice cream in this activity using different types of salts, but you may get different results. *How does ice cream made in a bag with table salt compare with ice cream made with rock salt or some other type of salt? Can you explain your results?* Note: Never eat rock salt. It could cause sodium poisoning.

OBSERVATIONS AND RESULTS

Did the ingredients that were shaken in the large bag with the ice cubes and salt turn into ice cream whereas the ingredients that were shaken with ice cubes alone (and no salt) remain fluid, and not harden into ice cream?

You should have seen that the ice cubes in the large bag with salt melted much more, and felt much colder, than the ice cubes in the large bag without salt. Because it was cold enough (several degrees below freezing), the ice cube bag with salt should have been able to cool the ingredients enough to harden them and turn them into ice cream whereas the ice cube bag without salt wasn't cold enough to do this, leaving the ingredients fluid. (If you put the fluid ingredients back in the bag with ice cubes and salt and shook them for about five minutes, the ingredients should have been cooled sufficiently to turn into ice cream!)

If you have ever made ice cream with an old-fashioned hand-crank machine, you probably packed a mixture of ice and rock salt around the container holding the cream. The ice-salt combination gets colder than pure water ice and can freeze the ingredients in the ice cream machine (and in the bags you used in this activity), turning them into ice cream. (This is the same process that occurs when icy roads have salt spread on them to melt the ice, keeping the roads less slippery at lower temperatures.) Although pure water freezes at 32°F (0°C), water mixed with salt will only freeze at colder temperatures below 32°F (0°C).

CLEANUP

If you haven't already, you can enjoy your tasty ice cream treat now or save it in the freezer for later.

Sweet Science

Making Marshmallows

HAVE YOU EVER WONDERED HOW MARSHMALLOWS ARE MADE?

Whether you're gathered around a fire or drinking hot chocolate after a day in the snow, nothing says sweet, squishy fun quite like a marshmallow! Even its name is soft and spongy! Long ago people made marshmallows with ingredients from the marshmallow herb, but today we usually make them with other ingredients, namely gelatin, corn syrup, and sugar. In this appetizing activity, you'll get to explore what ratio of sugar to corn syrup produces the best-tasting and best-textured marshmallows!

PROJECT TIME

45 to 60 minutes with follow up time within one week

KEY CONCEPTS

Chemistry
Boiling point
Food science

BACKGROUND

Marshmallows are an unusual type of sweet treat—spongy, sticky, and a little bit chewy. They have a melting point that is just above body temperature so that they start to change from a solid to a liquid state as soon as they reach the warmth of your mouth—or the heat of a fire! They're also an ancient creation, originally coming from a tall marshmallow plant (*Althaea officinalis*) that grows in swampy fields and has a soft, spongy root. Its root contains mucilage, a thick, gluey substance produced by some plants and microscopic animals to help with food storage and seed germination. Some cultures used the plant to make candy, whereas others used it to make medicine. The ancient Egyptians, for example, dried the root and mixed it with honey to make marshmallow treats, but the French experimented with using its gummy juice to soothe sore throats.

Modern marshmallows no longer contain parts of the marshmallow plant. Instead, the store-bought version is primarily a mix of three ingredients: sugar, corn syrup, and gelatin. The gelatin replaces the thick, gluey substance from the marshmallow plant. Varying the ratio of sugar to corn syrup can significantly affect what the resultant marshmallows are like.

MATERIALS

- Timer
- Two square foil cake pans or round foil pie pans each 8 to 9 inches (20 to 23 cm) wide
- Masking tape
- Pen or marker
- Vegetable oil, such as canola or safflower oil
- Paper towels
- Powdered or confectioner's sugar
- Small strainer
- Large mixing bowl
- Measuring cups
- Water
- Two plain, unflavored, 1/4-ounce (7 g) gelatin envelopes; available at grocery stores
- Fork
- 1/2 cup (118 ml) of corn syrup, like light corn syrup; glucose syrup is a possible substitute
- 1 1/4 cups of granulated sugar

- Small saucepan with lid
- Candy thermometer (must be able to read up to 240°F, or about 115°C)
- Electric mixer or beater
- Measuring spoons
- Pure vanilla extract
- Spatula
- Pizza wheel or cookie cutter approximately 1-by-1 inch (2.5-by-2.5 cm)
- Ruler
- Large airtight containers or gallon-size sealable plastic bags
- Taste-testing volunteers (such as friends and family!)

PREPARATION

- Wash and thoroughly rinse your hands.
- Set out all cooking tools and ingredients so they are ready to go and easy to access.
- Use the masking tape and a pen or marker to label the bottom of one cake pan "1" and the other "2."
- Pour a small amount of vegetable oil on a paper towel and lightly oil the inside of the cake pans. Pour a small amount of powdered sugar in the strainer and lightly dust the inside of the pans.
- Be careful when heating syrup mixtures in the saucepan. An adult should closely supervise these steps.

PROCEDURE

- Pour 1/6 cup (39 ml) of cold water into the large mixing bowl. (To get 1/6 cup, just fill up the 1/3 measuring cup halfway.) Sprinkle 1/4-ounce (7 g) packet of gelatin over the cold water. Mix the gelatin and water together for about five seconds with a fork and set the bowl aside. This will give the gelatin time to soften, or "bloom."

- Add 1/4 cup (59 ml) of cold water, 1/2 cup of sugar, and 1/3 cup (78 ml) of corn syrup to the saucepan. (This is a 3:2 ratio of sugar to corn syrup.)

- Put the lid on the saucepan and turn the stove to a medium-high heat. An adult should closely supervise all work from this point on. The syrup solution, which will become very hot, should be handled with extreme care.

- Lift the lid and check the solution in the saucepan about every 30 seconds until it just comes to a boil. Remove the lid.

- Begin to measure the temperature of the syrup in the saucepan using the candy thermometer. Do not let the thermometer touch the bottom or side of the pan, but instead try to put the tip below the surface, near the middle of the pan. Continue heating your syrup until the temperature reaches 240°F (about 115°C). *How does the temperature rise? Does it go up quickly at first? What happens as the syrup becomes more concentrated?*

- When the temperature reaches 240°F (about 115°C), immediately turn off the stove and move to the next step. *How does the syrup mixture seem to change as it gets closer to this temperature?*

- Turn the mixer on low and carefully pour the hot syrup into the large mixing bowl with the gelatin. (Some syrup will likely solidify on the saucepan.)

- Set the timer for 11 minutes and start it.

- Gradually increase the speed of the mixer until it is operating at full (high) speed. Continue to beat for approximately 11 minutes, or until the mixture starts to become very thick, glossy and lukewarm. *How does the mixture change as you mix it?*

- Add 1/2 teaspoon (2.5 ml) of vanilla extract and then beat for one more minute. Note how long you beat the syrup.

- Quickly use a vegetable oil-coated spatula to scoop out the marshmallow mixture from the mixing bowl and put it into the prepared cake pan labeled 1. If needed and easy to do, use the spatula to gently smooth down the top of the marshmallow so that it is pretty flat.
- Wash and dry all cooking utensils.
- Repeat the entire activity procedure (not including the preparation), but this time add 3/4 cup of sugar and 1/6 cup (39 ml) of corn syrup to the saucepan. (This is a 9:2 ratio of sugar to corn syrup.) (Still add the 1/4 cup [59 ml] of cold water to the saucepan first.) In the mixing bowl, beat this recipe for the same amount of time. When it is ready, scoop this marshmallow mixture from the mixing bowl and into the prepared cake pan labeled 2.
- Allow the marshmallow "pies" to sit out, uncovered, on a counter for at least four hours (or as long as overnight) so they can become firm.
- Once firm, turn the marshmallow pies over, one at a time, onto a cutting board. Use a spatula to help lift the pies out.
- Roll a pizza cutter or cookie cutter into some powdered sugar and then use it to cut up your marshmallows into approximately 1-by-1-inch (2.5-by-2.5 cm) pieces.
- Dust the marshmallows on all sides with a little strained powdered sugar.
- Place the marshmallows in an airtight container or plastic bag, labeling the container with the number of the recipe (1 or 2).
- Within a week of making the marshmallows, take a marshmallow piece from each bag. Try to find pieces that are as similar in shape and size as possible.

- Drop each marshmallow in a saucepan of hot water, keeping track of which marshmallow was from each recipe. Watch and see how quickly each marshmallow piece melts. Repeat this melting test three times. *Which marshmallow piece melted first?*
- Within a week of making the marshmallows, gather your taste testing volunteers (such as family and friends) to evaluate the marshmallows for taste and texture. *Which marshmallows do they prefer? Which ones are the chewiest and which ones are the least chewy? Which are the softest and which are the hardest?*

EXTRA

In this activity, you may have noticed that marshmallow pieces were harder from one recipe than the other. You can try to investigate this more quantitatively. To do this, put a marshmallow square from one of the recipes on a cutting board. Arrange a small cookie cutter so that it can sit on top of the square. On top of the cookie cutter balance a small plastic or paper cup. Put 20 pennies in the cup, one at a time. *Did the cookie cutter leave a noticeable impression?* Try this with a marshmallow square from the other recipe. *Did the cookie cutter leave a more or less noticeable impression?*

OBSERVATIONS AND RESULTS

Were the marshmallow pieces from recipe 1 fluffier, softer and less sweet than the pieces from recipe 2? Did the marshmallows from recipe 1 melt faster than the pieces from recipe 2?

In the sugar and corn syrup solutions you heated, the sugar made the solutions' boiling points higher than that of pure liquid water. The greater the concentration of sugar in the solution, the higher its boiling point. Consequently, marshmallow pieces from recipe 1 should have had a lower boiling point than pieces from recipe 2, making the former melt faster than the latter when placed in hot water. The final concentration of sugar in the syrup determines the structure of the candy that forms. Think about caramels and lollipops—the caramel is softer and chewier whereas the lollipop is hard and cracks when you bite it. The syrup used to make caramels is cooked until it has about an 87 percent concentration of sugar in solution, whereas the syrup used to make lollipops reaches about 99 percent concentration of sugar in solution. (As the syrup boils, water evaporates, and the syrup becomes even more concentrated with sugar.) When making marshmallows, the syrup is cooked until an 85 percent concentration of sugar is in the syrup, and the corn syrup is added to help prevent crystals from forming in the cooled syrup. The marshmallow pieces from recipe 1 should have been fluffier, softer, less sweet, and more like store-bought marshmallows than the pieces from recipe 2.

Gelatin is a protein that comes from collagen, the main protein in connective tissue in animals. It has the special quality of being able to coagulate (or come together) when it is beaten or whipped. When the hot concentrated syrup is beaten into gelatin with a mixer, bubbles form in the mixture and the gelatin coagulates around the bubbles, stabilizing their walls so they don't collapse.

CLEANUP

Enjoy your tasty marshmallow treats! If stored in an airtight container, they should be good for a week. To clean up any marshmallow-making messes, try dissolving and washing them away with warm water.

Find the Candy Rainbow with Chemistry

HERE IS A FUN PROJECT YOU CAN TRY WITH LEFTOVER CANDY YOU HAVE.

You will make some amazing art using nothing but colored candy and hot water—and a little bit of science. Save some of your candy and get started!

PROJECT TIME

30 to 40 minutes

KEY CONCEPTS

Chemistry
Molecules
Diffusion
Gradient
Solutions

BACKGROUND

Many of the foods we eat are colored with artificial dyes. Some of these dyes will easily dissolve in water. This means that when foods containing the dyes are placed in water, the molecules that make up the dye will break away from the food and start to spread out in the water. The process where molecules spread out like this is called diffusion. The molecules tend to spread from areas where there are a lot of that type of molecule (a high concentration) to areas where there are fewer of that type of molecule (a lower concentration). This is a gradient. Eventually the molecules all spread out enough that the concentration is equal everywhere. This is called equilibrium. The molecules, however, do not stop moving completely at that point. They will still bounce around randomly, and diffusion will still occur—but much more slowly.

Does all that sound confusing? Don't worry! You experience molecular diffusion every day with your sense of smell. Have you ever noticed how a smell will seem stronger when you are very close to the source or when the source of the smell just started out (for example, pulling a batch of fresh-baked cookies out of the oven)? This is because initially there is a higher concentration of odor-causing molecules in the air close to the source. Gradually these molecules will diffuse through the air into the rest of the room so you can detect the smell from farther away. In this test, you will see diffusion in action instead of smelling it—by using liquids instead of gases.

MATERIALS

- Hard-shelled colored candies, such as M&Ms or Skittles
- Small plate
- Glass or measuring cup
- Warm tap water
- Dish towel or paper towels
- Spoon
- Sugar
- Workstation than can tolerate spills and color dyes
- Food coloring (optional)

PREPARATION

- Arrange pieces of candy in a circle around the inner rim of the plate. Use at least two different colors, alternating them in groups of two or three.
- Fill a measuring cup or glass with warm tap water.

PROCEDURE

- Slowly pour warm tap water into the middle of the plate, until it partially covers the candy (or fully, depending on how deep the plate is).
- Watch the plate closely for a few minutes. *What happens?*
- Empty and dry off the plate.
- Make a circle of candy around the plate again.
- Put a small pile of sugar (about 1/4 teaspoon) directly in the middle of the plate.
- Slowly pour warm tap water near the center of the plate (but not directly onto the pile of sugar).
- Watch the plate closely for a few minutes. *What happens this time? Is it different from what happened the first time?*
- Clean off the plate and repeat the test. Try different color patterns and/or arranging the candy (and sugar) in different shapes. *What patterns and artwork can you make?*

SCIENCE FAIR IDEA

Watch the plate for about five to 10 minutes. *Do the colors continue to diffuse at the same rate they did initially?* Leave the plate out all day and check on it periodically or let it sit overnight. *How long does it take the colors to blend completely?*

SCIENCE FAIR IDEA

Try the test with cold water instead of warm water. *Do the colors diffuse at a different rate?*

EXTRA

Drop some food coloring into a plate of still water and watch how it diffuses. *How is it similar to or different from the way the dye diffused from the candy? Can you figure out why?*

OBSERVATIONS AND RESULTS

When you pour water onto the plate, the candy's colored coating starts to dissolve in the water. As a result you can see the colored dye diffuse towards the center of the plate. This diffusion can result in amazing, colorful rainbow patterns, as the colors remain mostly separate at first instead of bleeding together. You might have seen a "pie wedge" type pattern with different slices of each color you arranged around the rim.

When you put sugar in the middle of the plate, something strange happens. The dye seems to hit an invisible wall in the water and stops diffusing at first, then diffuses much more slowly. This occurs because in addition to colored dye the candy's coating also contains sugar. Both the sugar and the dye dissolve into the water, forming a mixture called a solution. Remember what you read in the background about gradients. With your first test the sugar wants to spread out from where there is a high concentration of sugar (right next to the candy) to where there is a lower concentration of sugar (the middle of the plate). The whole solution (sugar and dye molecules) moves along this gradient from high concentration to low concentration. Although you can't see the sugar from the candy coatings, you can see the colored dye. With your second test, however, there is already a lot of sugar in the middle of the plate—there is not a steep sugar gradient between the rim of the plate and the center. This prevents the sugar (and the dye) from spreading towards the center as rapidly. You might wonder: "Well, there are still no dye molecules in the middle of the plate, so shouldn't the dye molecules keep diffusing, even if the sugar doesn't?" That's a good question! Different types of molecules diffuse at different rates in water. In this case, the solution with the sugar molecules diffuses much faster than the dye molecules would on their own. This explains why, if you drop some food coloring into still water, it diffuses much more slowly than the dye from the candy. The food coloring does not contain any sugar.

If you watch the plate for a while, you might have noticed that the colors remain separated for quite some time. This can seem surprising—you might expect the colors to all blend together and turn a muddy brown. Remember, however, that it's the sugar gradient that drives the diffusion. After the whole plate is filled up with colors, the sugar concentration is the same everywhere—there is no sugar gradient between the different colors. So the colors will continue to diffuse slowly due to random motion of the molecules—but this process is much slower than the initial diffusion caused by the sugar gradient.

CLEANUP

You probably don't want to eat mushy candy, so dispose of any remaining candy or water in the trash. Use a towel to wipe up any spilled water.

THE SCIENTIFIC METHOD

The scientific method helps scientists—and students—gather facts to prove whether an idea is true. Using this method, scientists come up with ideas and then test those ideas by observing facts and drawing conclusions. You can use the scientific method to develop and test your own ideas!

Question: What do you want to learn? What problem needs to be solved? Be as specific as possible.
Research: Learn more about your topic and refine your question.
Hypothesis: Form an educated guess about what you think will answer your question. This allows you to make a prediction you can test.
Experiment: Create a test to learn if your hypothesis is correct. Limit the number of variables, or elements of the experiment that could change.
Analysis: Record your observations about the progress and results of your experiment. Then analyze your data to understand what it means.
Conclusion: Review all your data. Did the results of the experiment match the prediction? If so, your hypothesis was correct. If not, your hypothesis may need to be changed.

GLOSSARY

caramelize: To turn sugar into caramel by heating.
concentration: The amount of something in a certain space.
diffusion: The spreading or widely scattering of something.
fluorescent: Able to give off visible light.
fructose: A very sweet sugar found in honey and fruits.
germination: To cause to start to grow, sprout.
glucose: Sugar that occurs naturally in fruits, plants, and animals.
gradient: How much something is changed by an increase or decrease in pressure or temperature.
homogeneous: Having all parts the same.
interlace: Join together.
phenomenon: A fact or happening able to be known by the senses.
quantitative: Able to be measured.
solute: Something that has been dissolved.
texture: The feel or look of something.
transparent: Clear, easy to see through.
vigorously: Having an energetic or robust quality.

ADDITIONAL RESOURCES

Books

America's Test Kitchen. *The Complete Cookbook for Young Scientists.* Boston, MA: America's Test Kitchen, 2021.

Gates, Stefan. *Science You Can Eat: 20 Activities that Put Food Under the Microscope.* London, UK: DK Children, 2019.

LaFleur, Sara. *How Does My Body Work? Human Body Book for Kids: STEAM Experiments and Activities for Kids 8-12.* NewYork, NY: Z Kids, 2021.

Websites

Discovery Education
www.sciencefaircentral.com

Exploratorium
www.exploratorium.edu/search/science%20fair%20projects

Science Buddies
www.sciencebuddies.org/science-fair-projects/project-ideas/list

Science Fun for Everyone!
www.sciencefun.org/?s=science+fair

Videos

How to Cookie with Science/Reactions
ny.pbslearningmedia.org/resource/cookie-with-science-reactions/cookie-with-science-reactions/

Thermal Energy | DIY Science Time
bit.ly/3H2QOat

Publisher's note to educators and parents: Our editors have carefully reviewed these websites to ensure that they are suitable for students. Many websites change frequently, however, and we cannot guarantee that a site's future contents will continue to meet our high standards of quality and educational value. Be advised that students should be closely supervised whenever they access the internet.

INDEX